THE
ULTIMATE
PROMPTING
GUIDE

Your step-by-step guide to thinking smarter,
moving faster, and achieving more with AI

SCOTT DUFFY

THE ULTIMATE PROMPTING GUIDE

Your Step by Step Guide to Thinking Smarter, Moving Faster,
and Achieving More with AI

Scott Duffy

Copyright © 2026

Scott Duffy

ISBN: 978-1-968149-10-9

Joint Venture Publishing

The Millionaire Mentor, Inc

JOINT VENTURE
PUBLISHING

Printed in the United States of America

*"For those who dream big and dare
to ask the right questions."*

And

To Lily & Lexi
I am so proud to be your dad

*"You already have the answers.
You just need to learn how to ask the right
questions."*

— Scott Duffy

TABLE OF
CONTENTS

INTRODUCTION

Embracing AI as Your New Digital Assistant

Why Start with ChatGPT and How Do I Get Started Today?

The Art of Prompting & How to Use This Book

What You'll Learn (and Why It Matters)

01 SECTION
The Prompting Playbook:
How to Get World-Class Results with AI

Before you dive into the 100 ultimate prompts, you need one essential skill: learning how to talk to AI in a way that gets extraordinary results. This section gives you the exact framework I use to generate world-class output... every time.

02 SECTION
Business Prompts:
AI as Your Ultimate Growth Partner

The lifeblood of business is attention, and attention is won with the right words at the right time. These prompts turn ChatGPT into your personal copywriter, strategist, and creative department so that you can launch campaigns, grow audiences, and make your brand unforgettable.

Sales feed the business, but conversations spark the sale. These prompts will give you magnetic messages, persuasive pitches, and confidence to turn strangers into paying customers—on demand.

Without a plan, even great ideas can fail. These prompts turn AI into your strategic partner—helping you make bold, smart moves and see around corners.

Your product gets customers. Your experience keeps them. These prompts help you deliver moments so good, they'll talk about you for years.

Innovation is no longer optional; it's survival. These prompts will help you dream, design, and deliver products customers didn't even know they needed.

Businesses don't grow; people do. These prompts will help you lead with clarity, inspire action, and build teams that win together.

Numbers tell the story. These prompts will help you understand the plot and rewrite it in your favor.

In crowded markets, the most distinctive brand wins. These prompts help you shape an identity that can't be ignored.

Deals are won and lost in the pitch. These prompts will give you visuals, words, and stories that elicit a "yes."

Let technology handle the repetitive tasks so that you can focus on the visionary ones.

Get seen. Get heard. Get remembered.

One connection can change everything. These prompts help you make every interaction count.

03 SECTION
Personal Prompts:
AI as Your Life Guide, Coach & Spiritual Ally

Clarity is power. These prompts help you see yourself fully—and decide what comes next.

Without structure, focus, and momentum, dreams are just dreams. Use AI to help realize them.

Time is your most valuable currency. Spend it wisely.

Your body is your ultimate vehicle for success. Keep it running at peak performance.

You don't have to carry the weight alone. Let AI help you uncover strategies for better emotional health.

Strong relationships are built on strong communication. These prompts help you encourage both.

Whether you call it God, the Universe, or Love, the answers are always there. AI can help you listen.

Creativity is in our nature, but it doesn't always come naturally. These prompts will unlock what's waiting to be expressed.

Wealth starts in the mind long before it appears in the bank.

The more you grow, the more life flourishes.

Life's too short for "someday." Let's make it no w.

When you change the way you see the world, the world changes.

Your story is your gift to the future.

*Strong relationships are built on strong communication.
These prompts help you encourage both.*

*Whether you call it God, the Universe, or Love, the answers are
always there. AI can help you listen.*

*Creativity is in our nature, but it doesn't always come naturally.
These prompts will unlock what's waiting to be expressed.*

Wealth starts in the mind long before it appears in the bank.

The more you grow, the more life flourishes.

Life's too short for "someday." Let's make it no w.

When you change the way you see the world, the world changes.

Your story is your gift to the future.

INTRODUCTION

Embracing AI as Your New Digital Assistant

Throughout human history, we've encountered pivotal moments; inflection points that transformed our world. From the printing press to the Industrial Revolution, from the first automobiles to the advent of electricity, each leap forward brought new opportunities wrapped in a little bit of uncertainty. Today, we're on the brink of another monumental shift: the rise of AI.

What makes this era so exciting is that AI is incredibly accessible. It's a tool everyone can tap into. It's technology that is no longer just for the rich or tech elite, but for anyone ready to learn how to talk to it. Instead of something to fear, think of AI as a new kind of digital assistant that's right at your fingertips. One that can help both your business and personal life flourish.

This book is your roadmap to doing exactly that. We'll start by showing you how to get onto ChatGPT (if you haven't already), and then dive into the essentials: how to prompt like a pro. You'll learn the difference between prompting and just searching, how to craft the perfect prompts, and how to give AI the right roles, instructions, and context to really shine.

By the end of this guide, you'll not only have the know-how, but also a set of practical prompts for both your business and personal life.

Whether you're a CEO, student, creator, coach, or busy parent, these prompts will give you:

- More clarity (What should I do next? How do I think through this?)
- More speed (How can I get it done faster without sacrificing quality?)
- More leverage (How can I scale myself or my ideas?)
- More confidence (How do I know I'm on the right path?)
- Less overwhelm (How do I stop spinning my wheels?)

So let's get started on this journey together and turn AI into your ultimate resource.

Why Start with ChatGPT and How Do I Get Started Today?

As you embark on your AI journey, the perfect place to begin is with ChatGPT. Why ChatGPT? Because it's the most widely used and fastest-growing AI model in the world, now engaging over 100 million daily active users. Its massive user base and rapid integration into businesses make it the best platform to start with.

What's great is that the prompting skills you'll learn here aren't just for ChatGPT. They are universal. Whether you move on to use Google's Gemini, Anthropic's Claude, xAI's Grok, or any other model, the principles remain the same.

If you haven't signed up for ChatGPT yet, it's simple to get started. Begin with a free account. Go to www.chatgpt.com or download the ChatGPT app on your phone. You'll have full access to the latest models and all the tools you need. If you want to enhance your experience even further, there's a Pro plan available for around $20 a month, offering extra features and even more powerful capabilities.

So let's dive in and get you set up. Once you're on board, the rest of this book will guide you through mastering AI conversations that you can take anywhere.

The Art of Prompting & How to Use This Book

Every great conversation with AI starts with a prompt... the words you type or speak to tell it what you want.
Most people treat AI like Google: they ask short, vague questions and get generic, surface-level responses.

But that's not how you unlock AI's power. Real results happen when you learn to speak AI's language.

When you give AI a specific role, clear instructions, and rich context, something amazing happens: It stops sounding like a robot and starts thinking like an expert who knows you—your voice, your goals, your situation.

This is where the magic lives. It's not luck. It's a learnable skill. And in a world where AI is becoming the co-pilot for business and life, this skill will become your superpower.

What You'll Learn (and Why It Matters)

This book will teach you how to become fluent in the new language of prompting.

It's broken down into three powerful sections:

Section 1: The Prompting Playbook
You'll learn a simple but transformational 3-part framework—Role / Instruction / Context—for writing prompts that produce world-class results. We'll show you exactly how to go from "good enough" prompts to "elite-level" prompting that can 10X your creativity, clarity, and output.

Section 2: Business Prompts
Next, you'll get 50 proven prompts that help you grow your business, increase revenue, streamline operations, and scale your influence. We've organized them by department—like marketing, sales, leadership, and finance—so you can plug them directly into your day-to-day work.

Section 3: Personal Prompts
You'll get 50 more prompts to help you in your personal life. Whether it's setting goals, improving your health, unlocking creativity, healing emotionally, or even exploring spiritual growth, these prompts are here to support your full journey as a human being—not just a professional.

By the time you finish, you'll walk away with:

- A clear framework for getting AI to do exactly what you want
- 100 powerful prompts to guide your business and life

- The confidence to use AI as your personal growth partner

Because this isn't just about learning how to "use" AI.

It's about becoming superhuman... faster, smarter, more capable... by mastering the one skill that unlocks it all: the art of asking better questions.

So let's get started!

SECTION

01

THE PROMPTING PLAYBOOK

How to Get World-Class Results with AI

Before you dive into the 100 ultimate prompts, you need one essential skill: learning how to talk to AI in a way that gets extraordinary results. This section gives you the exact framework I use to generate world-class output... every time.

What Prompting Is (and What It's Not)

Most people treat ChatGPT like Google—typing short, vague questions and hoping for useful answers.

But AI isn't a search engine.

It's more like having a conversation with a PhD-level expert in every field imaginable. You can assign it any role—from marketing strategist to spiritual advisor—and it will tailor its answers to your exact situation.

If you give AI the right direction, it can:

- Think like the best minds in your industry.
- Spot patterns you might miss.
- Build plans, content, and strategies in seconds. Work that could take you weeks on your own.

Here's the key: **The quality of your answers depends entirely on the quality of your prompts.**

Prompting vs. Search: A Whole New Mindset

Most of us grew up "Googling it." We'd type a question, skim a few links, and piece together the information we needed.

Prompting is different.
You're not searching anymore. You're directing.

You're telling AI what to do and how to do it. It doesn't just retrieve info—it generates new insights, tailored just for you.

To tap into that power, you need a new mindset. If you approach prompting with openness and curiosity, you'll move faster than everyone else and unlock your superhuman edge.

The 3-Part Prompt Framework: Role / Instruction / Context

This framework is simple, but it will change how you use AI forever. Every prompt in this book uses it. Once you master it, you'll never again ask, "Why didn't ChatGPT give me what I wanted?"

👥 Step 1: Role
Who do you want AI to be in this conversation?

Instead of:
"Help me write a business plan."

Try:
"Your role is to act as the #1 business plan writer in the world with 20 years of experience writing plans that create multimillion dollar businesses."
Defining the Role sets perspective and unlocks expert-level responses.

✍️ Step 2: Instruction

What exactly do you want AI to do?

Don't say:

"Write me a business plan."

Do say:

"Write a business plan for a new fitness studio in Austin, Texas, including market analysis, target audience, unique selling proposition, pricing model, and 90-day marketing strategy."

Think of this as your to-do list for the AI. Be specific and step-by-step.

🌐 Step 3: Context

What background does AI need to make the output specific to you?

If Role is "who" and Instruction is "what," Context is the "why" and "how."

Feed it details like:

- Your product or service
- Your audience
- Goals, timelines, or budget
- Past wins, failures, or examples you admire

The more you give, the more personalized the response. Without context, AI is guessing.

Optional: Output

Once you've nailed Role, Instruction, and Context, you can define the Output format to make responses more actionable.

Examples:
- "Output your answer as a 5-slide PowerPoint outline."
- "Write this in the tone of Steve Jobs giving a keynote."
- "Provide a table with three columns: Step, Action, Deadline."

Good Prompt vs. Great Prompt

GOOD:
Write a social media post for my new product launch.

GREAT:
- **Role:** Act as the world's #1 direct-response copywriter, specializing in viral social media.
- **Instruction:** Write a high-converting post for our product launch, including a scroll-stopping hook, value prop, and CTA. Provide 3 variations.
- **Context:** The product is GreenFuel, a plant-based protein powder for busy professionals. It tastes better and has more protein than Vega or Garden of Life. We're launching online next month with a 20% discount for the first 500 buyers.

See the difference? The second prompt tells AI exactly what to do—and why.

How to Use the Ultimate Prompting Guide

The 100 prompts in this book aren't meant to be copy-pasted without thought. They're templates—starting points you can tailor using the **Role / Instruction / Context** format.

Treat them like search terms, and you'll get average answers.

Use the framework, and you'll get transformative results—the kind that can change your business, your life, and how you think.

Practice Prompt: Try It Now

Let's try your first custom prompt (be sure to customize it for your unique situation):

- **Role:** You are a personal productivity coach with 15 years of experience helping entrepreneurs design high-performance routines.
- **Instruction:** Create a custom morning routine that boosts my focus, energy, and productivity. Include wake-up time, exercise, nutrition, and mindset habits.
- **Context:** I'm a 42-year-old small business owner who works from home. I have two kids (ages 7 and 10), usually wake up at 6:30 AM, and feel sluggish in the mornings. I want more energy for my business and family and to lose 10 pounds in the next 3 months.

Fill in your own details. Run it in ChatGPT. Watch what happens.

SECTION

2

BUSINESS PROMPTS

AI as Your Ultimate Growth Partner

MARKETING & CONTENT CREATION

The lifeblood of business is attention, and attention is won with the right words at the right time. These prompts turn ChatGPT into your personal copywriter, strategist, and creative department so that you can launch campaigns, grow audiences, and make your brand unforgettable.

1. How do I create a high-converting marketing campaign for my product/service?

👥 Role:

Your role is to act as the world's #1 marketing strategist and copywriter, with deep expertise in consumer psychology, storytelling, and campaign design. You have a proven track record of creating marketing campaigns that generate multi-million-dollar results across multiple industries.

📝 Instruction:

Design a complete, high-converting marketing campaign for my product/ service. This should include:

- A clear campaign objective and success metrics
- An ideal customer profile
- A compelling offer and value proposition
- Key messages and storytelling hooks
- A multi-channel rollout plan (including social, email, and paid ads)
- A 30-day content calendar aligned with the campaign goals

🌐 Context:

- My company is named (Insert company name here).
- Here is what we do: (Insert short description of your product/service).
- Here is the market we serve: (Insert target audience description, demographics, and psychographics).
- Here is how our customers benefit: (Insert main benefits and outcomes customers experience).
- Here is an example of a past marketing campaign we've run or liked: (Insert example details or links).

- Here is our biggest competitor: (Insert competitor name).
- Here is a link to my website: (Insert URL).
- Here are links to competitive websites: (Insert URLs).
- Our budget for this campaign is (Insert budget range).
- Our desired launch date is (Insert launch date).

2 . What is the best content strategy to grow my brand on [social platform]?

👥 Role:

Your role is to act as the world's #1 content strategist, with expertise in brand growth and audience engagement on all major social platforms. You have a deep understanding of platform algorithms, storytelling, and content repurposing for maximum reach.

📝 Instruction:

Create a complete content strategy to grow my brand on (Insert specific social platform). Include:

- Optimal posting frequency and timing
- The best content formats for my audience on this platform
- A 90-day content plan with weekly themes
- Strategies for engagement and community building
- Suggested calls-to-action for each content type

🌐 Context:

- My company is named (Insert company name here).
- Here is what we do: (Insert short description of your product/service).
- Our target audience is (Insert detailed audience description).
- Our content goals are (Insert growth objectives, e.g., brand awareness, lead generation, sales).
- Here are examples of content we've posted or liked from others: (Insert examples or links).
- Our biggest competitors on this platform are (Insert competitor names/handles).
- Here is a link to our current profile: (Insert URL).

3. How do I write an irresistible sales page or landing page?

👥 Role:

Your role is to act as the world's #1 direct-response copywriter, with expertise in high-converting landing pages and sales copy that drive measurable action.

📝 Instruction:

Write a high-converting sales page for my product/service that includes:

- A compelling headline and subheadline
- A persuasive opening hook
- Key benefits and features in bullet form
- Testimonials or social proof (placeholders if not available)
- A strong guarantee (if applicable)
- A clear, urgent call-to-action

🌐 Context:

- My company is named (Insert company name here).
- Here is what we do: (Insert product/service description).
- Our target audience is (Insert detailed description).
- The offer we are promoting is (Insert product/service details, price, bonuses, deadlines).
- Here are examples of sales pages we like: (Insert examples/links).
- Our biggest competitors are (Insert competitor names/links).
- Our unique selling proposition is (Insert USP here).

4. Can you generate a 90-day social media posting calendar for my business?

👥 Role:
Your role is to act as the world's #1 social media strategist, with expertise in content planning, brand storytelling, and audience engagement.

📝 Instruction:
Create a detailed 90-day social media posting calendar for my business that includes:

- Daily post ideas and formats (e.g., reels, carousels, static images, stories, polls)
- Caption ideas with hooks and CTAs
- Recommended hashtags and keywords
- Key dates, holidays, or events relevant to my brand
- Engagement tips for each post type

🌐 Context:
- My company is named (Insert company name here).
- Here is what we do: (Insert product/service description).
- Our target audience is (Insert audience details).
- The platforms we want to focus on are (Insert platforms).
- Our content goals are (Insert growth or conversion goals).
- Here are examples of past posts that performed well: (Insert examples/links).
- Our competitors' best posts look like this: (Insert examples/links).

5. How do I write engaging email campaigns that convert?

👥 Role:
Your role is to act as the world's #1 email marketing strategist, with deep expertise in crafting high-converting email sequences that drive opens, clicks, and sales.

📝 Instruction:
Write a complete email campaign for my product/service that includes:

- A welcome or lead-nurture sequence (3–5 emails)
- Subject lines designed for high open rates
- Body copy optimized for readability and persuasion
- CTAs for each email that drive measurable results

🌐 Context:
- My company is named (Insert company name here).
- Here is what we do: (Insert product/service description).
- Our target audience is (Insert audience details).
- The offer we are promoting is (Insert product/service details).
- Our primary goal for this email campaign is (Insert goal—e.g., sales, event sign-ups, lead nurturing).
- Our current email marketing platform is (Insert platform name).
- Here are examples of email campaigns we admire: (Insert examples/ links).

6. What are 10 viral content ideas for my industry?

👥 Role:
Your role is to act as the world's #1 viral marketing expert, with expertise in content trends, audience psychology, and platform-specific virality triggers.

📝 Instruction:
Generate 10 creative, attention-grabbing content ideas tailored to my industry that have high potential to go viral. For each idea, include:

- The hook
- The format
- The target emotion it should trigger
- Suggested platform(s) for maximum reach

🌐 Context:
- My company is named (Insert company name here).
- Here is what we do: (Insert product/service description).
- Our target audience is (Insert audience details).
- Our main platform(s) for content distribution are (Insert platforms).
- Here are examples of past content that performed well for us: (Insert examples/links).
- Here are examples of viral content in our industry from competitors: (Insert examples/links).

7. How do I repurpose existing content for maximum reach?

👥 Role:
Your role is to act as the world's #1 content repurposing strategist, with expertise in transforming core content into multiple high-performing formats for different channels.

📝 Instruction:
Create a content repurposing plan for my existing material that includes:

- How to break down long-form content into multiple short-form pieces
- How to adapt content for different platforms while maintaining brand consistency
- A suggested publishing schedule
- Additional creative ways to reuse and refresh content for ongoing engagement

🌐 Context:
- My company is named (Insert company name here).
- Here is what we do: (Insert product/service description).
- Our target audience is (Insert audience details).
- The main type of content we currently produce is (Insert content type—e.g., blog posts, YouTube videos, podcasts).
- Here is a link to existing content we want to repurpose: (Insert URLs).
- Our primary platforms for distribution are (Insert platforms).
- Our goal for repurposing is (Insert goal—e.g., increased reach, lead generation, audience engagement).

SALES & LEAD GENERATION

Sales feed the business, but conversations spark the sale. These prompts will give you magnetic messages, persuasive pitches, and confidence to turn strangers into paying customers—on demand.

8. How do I write a cold email that gets responses from my ideal customers?

👥 Role:

Your role is to act as the world's #1 B2B/B2C sales copywriter and outreach strategist, specializing in cold email campaigns that get high open and reply rates.

📝 Instruction:

Write a cold email for my product/service that:

- Has a subject line designed for maximum opens
- Builds trust and credibility quickly
- Clearly states the problem we solve
- Includes a compelling offer or reason to reply
- Ends with a low-friction call-to-action

🌐 Context:

- My company is named (Insert company name here).
- Here is what we do: (Insert short description of product/service).
- Our target audience is (Insert detailed audience description).
- Our ideal customers are (Insert job titles, industries, or demographics).
- The main problem we solve for them is (Insert primary pain point).
- Here is the main benefit of working with us: (Insert main value proposition).
- Here is an example of a cold email we've used before: (Insert example if available).

9. What is the best follow-up sequence for leads who haven't replied?

👥 Role:
Your role is to act as the world's #1 sales follow-up strategist, with expertise in designing sequences that re-engage leads without being pushy.

📝 Instruction:
Create a follow-up sequence for leads who haven't replied to my initial outreach. Include:

- The number of follow-ups and timing for each
- The subject lines for each email/message
- The body copy that builds trust, addresses objections, and encourages response
- A final "breakup" message if there's still no reply

🌐 Context:
- My company is named (Insert company name here).
- Here is what we do: (Insert short description of product/service).
- Our target audience is (Insert detailed audience description).
- Our outreach platform is (Insert email, LinkedIn, or other tool).
- Our offer is (Insert brief offer details).
- Here is the tone we want to use: (Insert tone—e.g., professional, friendly, casual, urgent).

10. How do I write a persuasive sales pitch for [product/service]?

👥 Role:

Your role is to act as the world's #1 sales closer and presentation expert, specializing in creating pitches that convert high-value deals.

📝 Instruction:

Create a persuasive sales pitch script for (Insert product/service) that includes:

- A strong opening hook
- Key benefits and differentiators
- A relatable story or case study
- Responses to the top 3 objections
- A clear and compelling close

🌐 Context:

- My company is named (Insert company name here).
- Here is what we do: (Insert short description).
- Our target audience is (Insert detailed audience description).
- The main problem we solve is (Insert problem).
- The benefits of our product/service are (Insert top 3–5 benefits).
- Here is a link to our website: (Insert URL).
- Here are our main competitors: (Insert names and/or URLs).

11.Can you create a script for a discovery call with a new client?

👥 Role:
Your role is to act as the world's #1 sales coach, with expertise in structuring discovery calls that build rapport, qualify prospects, and set the stage for a close.

📝 Instruction:
Write a discovery call script for new prospects that includes:

- An opening to build rapport and set the agenda
- Questions to uncover needs, goals, and challenges
- Ways to position our offer as the ideal solution
- Transitional phrases that lead to next steps or a proposal

🌐 Context:
- My company is named (Insert company name here).
- Here is what we do: (Insert short description).
- Our target audience is (Insert details).
- The average deal size for our product/service is (Insert amount).
- The length of our sales cycle is (Insert average time).
- Our typical call length is (Insert time).

12. How do I identify and prioritize my most profitable target markets?

👥 Role:
Your role is to act as the world's #1 market research strategist, with expertise in identifying profitable customer segments and high-value market opportunities.

📝 Instruction:
Analyze my business and recommend my most profitable target markets, including:

- Primary audience segments by profitability and ease of reach
- Key demographics and psychographics for each segment
- Buying triggers and decision-making factors
- Recommendations for immediate outreach

🌐 Context:
- My company is named (Insert company name here).
- Here is what we do: (Insert product/service description).
- Our current customers are (Insert description).
- Our highest-value customers are (Insert description).
- Here are our top 3 competitors: (Insert names).
- Our geographic reach is (Insert regions or countries).

13. What is the best way to turn inbound inquiries into paying clients?

👥 Role:

Your role is to act as the world's #1 market research strategist, with expertise in identifying profitable customer segments and high-value market opportunities.

📝 Instruction:

Analyze my business and recommend my most profitable target markets, including:

- Primary audience segments by profitability and ease of reach
- Key demographics and psychographics for each segment
- Buying triggers and decision-making factors
- Recommendations for immediate outreach

🌐 Context:

- My company is named (Insert company name here).
- Here is what we do: (Insert product/service description).
- Our current customers are (Insert description).
- Our highest-value customers are (Insert description).
- Here are our top 3 competitors: (Insert names).
- Our geographic reach is (Insert regions or countries).

STRATEGY & BUSINESS PLANNING

www.buddhiwisdomhub.com

Without a plan, even great ideas can fail. These prompts turn AI into your strategic partner—helping you make bold, smart moves and see around corners.

14. How do I create a one-page strategic plan for my company?

👥 Role:
Your role is to act as the world's #1 business strategist, with expertise in creating clear, actionable strategic plans that align teams and drive growth.

📝 Instruction:
Create a one-page strategic plan for my company that includes:

- Our mission and vision statements 3–5 key strategic objectives
- Measurable goals and KPIs for each objective
- Action steps and responsibilities
- A clear timeline for execution

🌐 Context:
- My company is named (Insert company name here).
- Here is what we do: (Insert short description of your product/service).
- Our target audience is (Insert audience description).
- Our current business goals are (Insert 3–5 main goals).
- Our biggest challenges are (Insert challenges).
- Our team size and key roles are (Insert details).

15. Can you analyze my business model and suggest improvements?

👥 Role:
Your role is to act as the world's #1 business model optimization expert, with experience in scaling companies across multiple industries.

📝 Instruction:
Analyze my current business model and suggest at least 5 improvements that will increase profitability, scalability, and customer retention. Include:

- An overview of current strengths and weaknesses
- Opportunities for revenue growth
- Cost-saving or efficiency measures
- New channels or markets to explore

🌐 Context:
- My company is named (Insert company name here).
- Here is what we do: (Insert description).
- Our current revenue model is (Insert model—e.g., subscription, one-time sales, freemium).
- Our primary revenue streams are (Insert streams).
- Our current customer acquisition process is (Insert process).
- Here are our main competitors: (Insert names/links).

16. What is the best way to launch a new product in my industry?

👥 Role:
Your role is to act as the world's #1 business model optimization expert, with experience in scaling companies across multiple industries.

📝 Instruction:
Analyze my current business model and suggest at least 5 improvements that will increase profitability, scalability, and customer retention. Include:

- An overview of current strengths and weaknesses
- Opportunities for revenue growth
- Cost-saving or efficiency measures
- New channels or markets to explore

🌐 Context:
- My company is named (Insert company name here).
- Here is what we do: (Insert description).
- Our current revenue model is (Insert model—e.g., subscription, one-time sales, freemium).
- Our primary revenue streams are (Insert streams).
- Our current customer acquisition process is (Insert process).
- Here are our main competitors: (Insert names/links).

17. How do I conduct a SWOT analysis for my business?

👥 Role:
Your role is to act as the world's #1 business analyst, with expertise in identifying strategic strengths, weaknesses, opportunities, and threats for companies.

📝 Instruction:
Conduct a full SWOT analysis for my business, including:

- Strengths (internal advantages)
- Weaknesses (internal challenges)
- Opportunities (external growth possibilities)
- Threats (external risks and competition)
- Then, provide 3–5 recommendations for leveraging strengths and opportunities while addressing weaknesses and threats.

🌐 Context:
- My company is named (Insert company name here).
- Here is what we do: (Insert description).
- Our target audience is (Insert details).
- Our current market position is (Insert position—e.g., market leader, emerging player).
- Our biggest competitors are (Insert names/links).
- Our primary growth goals are (Insert goals).

18. How can I recession-proof my business for the next 12 months?

👥 Role:
Your role is to act as the world's #1 business resilience strategist, with expertise in helping companies thrive during economic downturns.

📝 Instruction:
Create a recession-proofing plan for my business that includes:

- Strategies to reduce unnecessary expenses
- Ways to diversify revenue streams
- Tactics to strengthen customer loyalty and retention
- Opportunities to negotiate better supplier/vendor terms
- Low-cost marketing strategies to maintain visibility

🌐 Context:
- My company is named (Insert company name here).
- Here is what we do: (Insert description).
- Our target audience is (Insert details).
- Our current revenue streams are (Insert streams).
- Our top 3 expenses are (Insert expenses).
- Here are our main competitors: (Insert names/links).

19. What are 5 new revenue streams I could add to my business?

👥 Role:

Your role is to act as the world's #1 business growth consultant, specializing in creative revenue generation strategies.

📝 Instruction:

Recommend 5 new revenue streams for my business that align with my brand, audience, and resources. For each, include:

- A description of the opportunity
- How it fits with my existing business model
- Estimated setup time and cost
- Potential monthly/annual revenue

🌐 Context:

- My company is named (Insert company name here).
- Here is what we do: (Insert description).
- Our target audience is (Insert details).
- Our current revenue streams are (Insert streams).
- Our available budget for new initiatives is (Insert budget).
- Our main competitors are (Insert names/links).

CUSTOMER SERVICE & EXPERIENCE

Your product gets customers. Your experience keeps them. These prompts help you deliver moments so good, they'll talk about you for years.

20. How do I design a customer onboarding process that delights and retains clients?

👥 Role:
Your role is to act as the world's #1 customer experience strategist, specializing in designing onboarding processes that create loyalty, increase retention, and turn customers into brand advocates.

📝 Instruction:
Design a customer onboarding process for my business that includes:

- Step-by-step touchpoints from purchase to first success
- Automated communications (emails, videos, or messages) to guide customers
- Personalization strategies for different customer segments
- Metrics to track onboarding success and engagement

🌐 Context:
- My company is named (Insert company name here).
- Here is what we do: (Insert short description of your product/service).
- Our target audience is (Insert audience description).
- Our current onboarding process is (Insert details, if any).
- The average time it takes for a customer to see value is (Insert time frame).
- Here is an example of onboarding we admire: (Insert link or description).

21. Can you create templates for responding to common customer complaints?

👥 Role:

Your role is to act as the world's #1 customer service manager, with expertise in empathetic communication, conflict resolution, and customer retention.

📝 Instruction:

Create a set of ready-to-use response templates for our most common customer complaints. For each template, include:

- A clear, empathetic opening
- A resolution or next step
- Language that reinforces trust and loyalty
- An offer or goodwill gesture (if appropriate)

🌐 Context:

- My company is named (Insert company name here).
- Here is what we do: (Insert product/service description).
- Our target audience is (Insert details).
- The top 3 customer complaints we receive are: (Insert complaints).
- Our brand voice and tone are (Insert tone—e.g., friendly, formal, casual).
- Here are examples of how we've responded in the past (Insert examples).

22. How do I write a proactive customer success outreach email?

👥 Role:

Your role is to act as the world's #1 customer service manager, with expertise in empathetic communication, conflict resolution, and customer retention.

📝 Instruction:

Create a set of ready-to-use response templates for our most common customer complaints. For each template, include:

- A clear, empathetic opening
- A resolution or next step
- Language that reinforces trust and loyalty
- An offer or goodwill gesture (if appropriate)

🌐 Context:

- My company is named (Insert company name here).
- Here is what we do: (Insert product/service description).
- Our target audience is (Insert details).
- The top 3 customer complaints we receive are: (Insert complaints).
- Our brand voice and tone are (Insert tone—e.g., friendly, formal, casual).
- Here are examples of how we've responded in the past: (Insert examples).

23. What are the best ways to increase customer loyalty and referrals?

👥 Role:
Your role is to act as the world's #1 customer loyalty and retention expert, with experience designing programs that turn happy customers into repeat buyers and brand ambassadors.

📝 Instruction:
Recommend a customer loyalty and referral strategy for my business that includes:

- Ways to reward repeat customers
- Referral program structure and incentives
- Ongoing engagement tactics to keep customers connected
- Metrics to measure loyalty and referral success

🌐 Context:
- My company is named (Insert company name here).
- Here is what we do: (Insert product/service description).
- Our target audience is (Insert details).
- Our current repeat purchase rate is (Insert percentage).
- We have/do not have a current referral program (Insert yes or no).
- Our biggest competitor's loyalty/referral program is (Insert link or description).

PRODUCT DEVELOPMENT & INNOVATION

Innovation is no longer optional; it's survival. These prompts will help you dream, design, and deliver products customers didn't even know they needed.

24. How do I validate a new product idea before launching?

👥 Role:

Your role is to act as the world's #1 product development strategist, with expertise in validating ideas through market research, prototyping, and customer feedback before launch.

📝 Instruction:

Create a product validation plan for my idea that includes:

- Steps to research demand and competition
- Methods to gather feedback from target customers
- MVP (minimum viable product) development guidelines
- Key metrics to determine whether to proceed or pivot

🌐 Context:

- My company is named (Insert company name here).
- Here is what we do: (Insert short description).
- The new product idea is (Insert description of idea).
- Our target audience is (Insert details).
- Our budget for validation is (Insert budget).
- Here is the timeline we want to work within: (Insert time frame).
- Our main competitors for this idea are (Insert names/links).

25. Can you create a feature roadmap for my SaaS product?

👥 Role:
Your role is to act as the world's #1 SaaS product manager, with expertise in prioritizing features based on customer needs, market trends, and business goals.

📝 Instruction:
Create a prioritized feature roadmap for my SaaS product that includes:

- High-priority features for immediate development
- Medium-term features to increase value and retention
- Long-term features for competitive advantage
- Suggested timeline for release phases

🌐 Context:
- My company is named (Insert company name here).
- Our SaaS product is called (Insert product name).
- Here is what it does: (Insert description).
- Our target audience is (Insert details).
- Our current feature set includes (Insert list).
- Our biggest customer requests are (Insert requests).
- Our main competitors and their key features are (Insert names/features).

26. How do I conduct a competitive analysis for my product?

👥 Role:
Your role is to act as the world's #1 competitive intelligence analyst, with expertise in identifying competitor strengths, weaknesses, and market positioning.

📝 Instruction:
Create a competitive analysis for my product that includes:

- A comparison table of my product vs. top competitors
- Key differentiators for each product
- Competitor pricing, positioning, and marketing strategies
- Opportunities for me to stand out in the market

🌐 Context:
- My company is named (Insert company name here).
- Here is what we do: (Insert description).
- Our product is (Insert name and description).
- Our target audience is (Insert details).
- Our main competitors are (Insert names/links).
- Here are features or benefits that make us unique: (Insert list).

27. What is the best way to collect and use customer feedback for improvements?

👥 Role:
Your role is to act as the world's #1 customer feedback strategist, with expertise in designing feedback systems that drive product improvement and customer loyalty.

📝 Instruction:
Recommend a system for collecting and using customer feedback that includes:

- Channels for gathering feedback (surveys, interviews, reviews, etc.)
- How to analyze and prioritize feedback
- How to close the loop with customers so they feel heard
- A process for implementing changes based on feedback

🌐 Context:
- My company is named (Insert company name here).
- Here is what we do: (Insert description).
- Our target audience is (Insert details).
- We currently gather feedback through (Insert methods).
- Our biggest challenge with feedback is (Insert challenge).
- Here is an example of feedback we've received recently: (Insert example).

LEADERSHIP & TEAM MANAGEMENT

Businesses don't grow; people do. These prompts will help you lead with clarity, inspire action, and build teams that win together.

28. How do I create a clear role description and responsibilities for my team members?

👥 Role:
Your role is to act as the world's #1 organizational development consultant, with expertise in designing clear, actionable role descriptions that align with business goals and set employees up for success.

📝 Instruction:
Create a detailed role description for (Insert role title) that includes:

- Role purpose and overall mission
- Key responsibilities and daily tasks
- Required skills, experience, and qualifications
- Performance metrics and expectations
- How this role supports the company's larger goals

🌐 Context:
- My company is named (Insert company name here).
- Here is what we do: (Insert short description).
- The department this role is in is (Insert department).
- The role title is (Insert title).
- The main purpose of this role is (Insert purpose).
- The skills and experience we need are (Insert list).
- The person in this role will report to (Insert manager/supervisor name or title).

29. Can you help me design a performance review system?

👥 Role:
Your role is to act as the world's #1 HR strategist, with expertise in creating performance review systems that motivate employees, encourage growth, and align with company objectives.

📝 Instruction:
Design a performance review system for my company that includes:

- Review frequency and schedule
- Core competencies and skills to evaluate
- A rating scale or scoring system
- Guidelines for goal-setting and development plans
- A process for follow-up and continuous improvement

🌐 Context:
- My company is named (Insert company name here).
- Here is what we do: (Insert description).
- Our team size is (Insert number of employees).
- Our company values and culture are (Insert description).
- Our current performance review process (if any) is (Insert details).
- The outcomes we want from performance reviews are (Insert outcomes).

30. What is the best way to handle conflict between team members?

👥 Role:
Your role is to act as the world's #1 workplace conflict resolution expert, with expertise in mediation, communication, and maintaining a positive team culture.

📝 Instruction:
Create a conflict resolution process for my company that includes:

- How to identify issues early
- A step-by-step mediation approach
- Guidelines for private and group conversations
- Methods for documenting conflicts and resolutions
- Ways to rebuild trust after a conflict

🌐 Context:
- My company is named (Insert company name here).
- Here is what we do: (Insert description).
- Our team size is (Insert number of employees).
- Our current conflict resolution process (if any) is (Insert details).
- Here is an example of a recent conflict: (optional) (Insert example).
- Our overall company culture is (Insert description).

31. How do I motivate a remote or hybrid team effectively?

👥 Role:
Your role is to act as the world's #1 remote team leadership coach, with expertise in keeping distributed teams engaged, productive, and connected.

📝 Instruction:
Recommend a remote/hybrid team motivation plan that includes:

- Regular communication and collaboration rituals
- Recognition and reward strategies
- Ways to build personal connections across locations
- T ools and technology to support engagement
- Metrics to track team morale and performance

🌐 Context:
- My company is named (Insert company name here).
- Here is what we do: (Insert description).
- Our team size is (Insert number of employees).
- Our team structure is (Insert fully remote, hybrid, etc.).
- Our current tools and platforms are (Insert tools).
- Our biggest challenge with remote/hybrid work is (Insert challenge).

FINANCIAL OPTIMIZATION

Numbers tell the story. These prompts will help you understand the plot and rewrite it in your favor.

32. How do I create a financial forecast for the next 12 months?

👥 Role:

Your role is to act as the world's #1 financial planning expert, with deep experience in building accurate, actionable forecasts for businesses of all sizes.

📝 Instruction:

Create a 12-month financial forecast for my company that includes:

- Projected revenue by month
- Projected expenses by month
- Profit and loss projections
- Cash flow projections
- Key assumptions used in the forecast

🌐 Context:

- My company is named (Insert company name here).
- Here is what we do: (Insert short description).
- Our current monthly revenue is (Insert amount).
- Our current monthly expenses are (Insert amount).
- Our growth goals for the next 12 months are (Insert goals).
- Our typical seasonal or market fluctuations are (Insert details).
- We operate in the (Insert industry) industry.

33. What is the best way to set and track KPIs for my business?

👥 Role:

Your role is to act as the world's #1 performance metrics consultant, with expertise in identifying and tracking KPIs that drive business growth.

📝 Instruction:

Identify the most important KPIs for my business and create a system to track them that includes:

- 5–10 KPIs aligned with business goals
- The formula for calculating each KPI
- Recommended frequency of measurement
- Tools or dashboards to monitor progress
- Suggested benchmarks or targets

🌐 Context:

- My company is named (Insert company name here).
- Here is what we do: (Insert description).
- Our short-term business goals are (Insert goals).
- Our long-term business goals are (Insert goals).
- Our team size is (Insert number).
- Our current performance tracking system is (Insert details or "none").

34. How do I identify unnecessary expenses and cut costs strategically?

👥 Role:
Your role is to act as the world's #1 business efficiency consultant, with expertise in cost reduction strategies that maintain or improve quality.

📝 Instruction:
Review my business operations and recommend cost-cutting opportunities that include:

- Expenses that can be reduced or eliminated
- Vendor or supplier renegotiation opportunities
- Process efficiencies to lower labor costs
- Technology or automation tools that save money
- The projected savings from each recommendation

🌐 Context:
- My company is named (Insert company name here).
- Here is what we do: (Insert description).
- Our current monthly expenses are (Insert amount).
- Our biggest expense categories are (Insert categories).
- Our cost reduction goal is (Insert amount or percentage).
- We have/do not have contracts with vendors; (Insert details).

35. Can you create a budget template for my business?

👥 Role:
Your role is to act as the world's #1 business efficiency consultant, with expertise in cost reduction strategies that maintain or improve quality.

📝 Instruction:
Review my business operations and recommend cost-cutting opportunities that include:

- Expenses that can be reduced or eliminated
- Vendor or supplier renegotiation opportunities
- Process efficiencies to lower labor costs
- Technology or automation tools that save money
- The projected savings from each recommendation

🌐 Context:
- My company is named (Insert company name here).
- Here is what we do: (Insert description).
- Our current monthly expenses are (Insert amount).
- Our biggest expense categories are (Insert categories).
- Our cost reduction goal is (Insert amount or percentage).
- We have/do not have contracts with vendors; (Insert details).

BRANDING & POSITIONING

In crowded markets, the most distinctive brand wins. These prompts help you shape an identity that can't be ignored.

36. How do I create a brand voice and style guide for my company?

👥 Role:
Your role is to act as the world's #1 brand strategist, with expertise in creating brand voice and style guides that ensure consistency across all marketing and communications.

📝 Instruction:
Create a brand voice and style guide for my company that includes:

- Brand personality traits and tone of voice
- Key messaging pillars
- Writing style guidelines (grammar, formatting, preferred terms)
- Visual style guidelines (color palette, fonts, image style)
- Examples of "on-brand" vs. "off-brand" communication

🌐 Context:
- My company is named (Insert company name here).
- Here is what we do: (Insert short description).
- Our target audience is (Insert details).
- Our current brand personality is (Insert traits—e.g., professional, playful, innovative).
- Here are examples of marketing materials we've used: (Insert links or descriptions).
- Here are competitors whose branding we like/dislike: (Insert links or names).

37. *What is the best tagline or slogan for my business?*

👥 Role:
Your role is to act as the world's #1 copywriter and brand messaging expert, with experience creating memorable taglines and slogans that capture a brand's essence.

📝 Instruction:
Create 10 tagline or slogan options for my business that:

- Clearly communicate our value proposition
- Are short, memorable, and easy to say
- Evoke the right emotion in our target audience
- Differentiate us from competitors

🌐 Context:
- My company is named (Insert company name here).
- Here is what we do :(Insert short description).
- Our target audience is (Insert details).
- Our unique selling proposition is (Insert USP).
- Here are examples of taglines we like: (Insert examples).
- Our brand personality is (Insert traits—e.g., bold, friendly, luxurious).

38. How do I position my brand as the go-to expert in my niche?

👥 Role:
Your role is to act as the world's #1 brand positioning consultant, with expertise in authority building, thought leadership, and niche domination strategies.

📝 Instruction:
Create a brand positioning plan for my company that includes:

- A clear positioning statement
- Content marketing strategies to showcase expertise
- PR and media outreach opportunities
- Partnership or collaboration ideas to boost credibility
- Metrics to measure brand authority growth

🌐 Context:
- My company is named (Insert company name here).
- Here is what we do: (Insert short description).
- Our target audience is (Insert details).
- Our biggest competitors are (Insert names).
- Our main expertise or differentiator is (Insert details).
- Here is our current positioning in the market: (Insert details).

39. What is the best way to refresh my brand without losing existing customers?

👥 Role:
Your role is to act as the world's #1 rebranding strategist, with expertise in evolving brands while maintaining loyalty from current customers.

📝 Instruction:
Create a rebranding plan for my company that includes:

- Which brand elements to keep vs. update
- How to communicate the change to existing customers
- Marketing campaigns to generate excitement about the refresh
- Timeline and rollout strategy
- Ways to measure rebrand success

🌐 Context:
- My company is named (Insert company name here).
- Here is what we do: (Insert short description).
- Our target audience is (Insert details).
- The reason for the rebrand is (Insert reason).
- Here are elements of our current brand we want to keep: (Insert details).
- Here are elements we want to change: (Insert details).
- Our biggest competitor's branding looks like this: (Insert description/ link).

PRESENTATIONS & PROPOSALS

Deals are won and lost in the pitch. These prompts will give you visuals, words, and stories that elicit a "yes."

40. How do I design a pitch deck for investors or partners?

👥 Role:

Your role is to act as the world's #1 pitch deck strategist, with expertise in crafting persuasive, visually compelling presentations that secure funding and partnerships.

📝 Instruction:

Create a complete pitch deck outline for my business that includes:

- Company overview
- Problem and market opportunity
- Solution and product/service details
- Business model and revenue streams
- Market analysis and competitive landscape
- Traction and key milestones
- Inancial projections
- Team bios and expertise
- Call-to-action for investors or partners

🌐 Context:

- My company is named (Insert company name here).
- Here is what we do: (Insert short description).
- Our target investors/partners are (Insert details).
- The funding/partnership amount we're seeking is (Insert amount).
- Our current traction includes (Insert key metrics, milestones, or wins).
- Our main competitors are (Insert names/links).
- Our pitch deck style preference is (Insert style—e.g., clean and minimalist, bold and visual).

41. Can you write a winning proposal for a major client?

👥 Role:

Your role is to act as the world's #1 business proposal writer, with expertise in winning high-value contracts and creating persuasive, client-focused proposals.

📝 Instruction:

Write a proposal for (Insert client name or type) that includes:

- A strong executive summary
- Clear understanding of the client's needs
- Our proposed solution and deliverables
- Timeline and milestones
- Pricing and payment terms
- Case studies or proof of past results\
- Closing statement and call-to-action

🌐 Context:

- My company is named (Insert company name here).
- Here is what we do: (Insert short description).
- The client we are targeting is (Insert details).
- The client's main needs are (Insert description).
- Our proposed solution is (Insert description).
- Our past work relevant to this client includes (Insert examples).
- Our budget/price for this proposal is (Insert price or range).

42. How do I make my presentations more persuasive and engaging?

👥 Role:

Your role is to act as the world's #1 presentation coach, with expertise in storytelling, persuasion, and audience engagement techniques.

📝 Instruction:

Review my presentation and recommend improvements that include:

- Stronger opening and closing statements
- Clearer structure and flow
- More persuasive storytelling elements
- Engaging visuals and multimedia
- Audience interaction techniques
- Strategies for delivering with confidence

🌐 Context:

- My company is named (Insert company name here).
- Here is what we do: (Insert short description).
- The type of presentation is (Insert type—e.g., sales pitch, keynote, training).
- T he audience for this presentation is (Insert audience details).
- Our current presentation length is (Insert duration).
- Here is a link to or description of the current presentation: (Insert link or description).

AI & AUTOMATION
FOR BUSINESS

Let technology handle the repetitive tasks so that you can focus on the visionary ones.

43. How can I use AI to streamline my daily business operations?

👥 Role:

Your role is to act as the world's #1 AI business operations consultant, with expertise in identifying, implementing, and optimizing AI tools to save time, reduce costs, and improve productivity.

📝 Instruction:

Recommend ways I can use AI to streamline my daily business operations, including:

- Specific AI tools or platforms to implement
- Tasks or processes to automate
- Steps for integrating these tools into existing workflows
- Estimated time and cost savings
- A 30-day action plan to get started

🌐 Context:

- My company is named (Insert company name here).
- Here is what we do: (Insert short description).
- Our team size is (Insert number of employees).
- The departments we want to improve with AI are (Insert list—e.g., marketing, sales, operations, finance).
- Our current tech stack includes (Insert tools and software used).
- Our biggest operational bottlenecks are (Insert issues).

44. *What are the top AI tools for marketing, sales, and productivity?*

👪 Role:

Your role is to act as the world's #1 AI tools and technology analyst, with expertise in evaluating software solutions for marketing, sales, and productivity.

📝 Instruction:

Provide a list of the top AI tools for marketing, sales, and productivity, including:

- Tool name and brief description
- Key features and benefits
- Pricing overview
- Ideal use cases
- Why it stands out compared to competitors

🌐 Context:

- My company is named (Insert company name here).
- Here is what we do: (Insert short description).
- Our main business priorities right now are (Insert priorities—e.g., lead generation, customer retention, scaling operations).
- Our current tools and platforms are (Insert list).
- Our monthly budget for AI tools is (Insert budget).

45. Can you create a workflow to automate repetitive tasks in my company?

👥 Role:
Your role is to act as the world's #1 workflow automation expert, with experience in designing and implementing systems that eliminate repetitive manual work.

📝 Instruction:
Create a detailed automation workflow for my company that includes:

- A list of repetitive tasks to automate
- Recommended tools or platforms for each task
- Step-by-step implementation plan
- Integration tips for connecting different systems
- Time and cost savings estimate

🌐 Context:
- My company is named (Insert company name here).
- Here is what we do: (Insert short description).
- Our team size is (Insert number of employees).
- The repetitive tasks we want to automate are (Insert tasks).
- Our current tech stack is (Insert tools/platforms used).
- Our budget for automation is (Insert budget).

PR & MEDIA OUTREACH

Get seen. Get heard. Get remembered.

46. How do I write a press release for a product launch?

👥 Role:

Your role is to act as the world's #1 PR and media communications expert, with decades of experience writing press releases that capture media attention and generate buzz.

📝 Instruction:

Write a professional press release for my product launch that includes:

- An attention-grabbing headline and subheadline
- A compelling opening paragraph with the key announcement details
- A clear description of the product/service and its benefits
- Relevant quotes from company leadership or experts
- Company background (boilerplate)
- Media contact information

🌐 Context:

- My company is named (Insert company name here).
- Here is what we do: (Insert short description).
- The product/service we are launching is (Insert description).
- Our target audience is (Insert details).
- Our launch date is (Insert date).
- Here is our company boilerplate: (Insert boilerplate).
- Our primary media contact is (Insert name, email, phone).

47. What is the best way to pitch myself to podcasts or media outlets?

👥 Role:
Your role is to act as the world's #1 media outreach strategist, with expertise in securing high-value press and podcast interviews for thought leaders, founders, and experts.

📝 Instruction:
Create a podcast/media pitch for me that includes:

- A short, engaging personal bio
- My credibility and expertise in the industry
- A unique hook or angle that will interest the audience
- 3–5 suggested interview topics
- A clear call-to-action for booking me as a guest

🌐 Context:
- My name is (Insert your name).
- My company is named (Insert company name here).
- Here is what we do: (Insert short description).
- My expertise is in (Insert details).
- My target media outlets/podcasts are (Insert names).
- Here are examples of past interviews I've done: (Insert links, if any).

48. How do I build relationships with journalists in my industry?

👥 Role:
Your role is to act as the world's #1 media relations expert, specializing in building authentic, long-term relationships with journalists and editors.

📝 Instruction:
Create a relationship-building plan for connecting with journalists in my industry that includes:

- How to identify the right journalists and publications
- Tips for engaging with them before making a pitch
- Value-first communication strategies
- How often to follow up without being pushy
- Ways to provide ongoing value and stay top of mind

🌐 Context:
- My company is named (Insert company name here).
- Here is what we do: (Insert short description).
- Our industry is (Insert industry).
- The publications/media outlets we want to target are (Insert names).
- Our goal for building journalist relationships is (Insert goal—e.g., more coverage, thought leadership, industry visibility).

EVENTS & NETWORKING

One connection can change everything. These prompts help you make every interaction count.

49. How do I plan and promote a successful business event?

👥 Role:
Your role is to act as the world's #1 event marketing strategist, with expertise in planning and promoting high-impact business events that drive attendance, engagement, and measurable ROI.

📝 Instruction:
Create a complete plan for my business event that includes:

- Event concept and theme
- Target audience and attendee goals
- Venue or platform recommendations (for in-person or virtual)
- Promotional timeline and channels
- Partnership or sponsorship opportunities
- Engagement strategies during the event
- Post-event follow-up plan

🌐 Context:
- My company is named (Insert company name here).
- Here is what we do: (Insert short description).
- The event we are hosting is (Insert event type—e.g., conference, workshop, networking mixer).
- The target audience is (Insert details).
- The planned date is (Insert date).
- Our budget for the event is (Insert budget).
- Our event location or platform is (Insert details).
- Our main competitors' events look like this: (Insert description or links).

50. What is the best way to network effectively at industry conferences?

👥 Role:
Your role is to act as the world's #1 business networking coach, with expertise in helping professionals maximize opportunities at conferences, trade shows, and industry events.

📝 Instruction:
Create a networking strategy for an upcoming conference that includes:

- How to research and identify high-value contacts before the event
- Conversation starters and value-based introductions
- How to position myself and my business in conversations
- Tactics for following up effectively after the event
- Ways to turn casual connections into long-term relationships

🌐 Context:
- My name is (Insert your name).
- My company is named (Insert company name here).
- Here is what we do: (Insert short description).
- The conference I'm attending is (Insert name of conference).
- The date/location of the event is (Insert details).
- The type of people I want to meet are (Insert target roles/industries).
- My networking goals for the event are (Insert goals—e.g., generate leads, find partners, raise awareness).

SECTION

03

PERSONAL PROMPTS

AI as Your Life Guide, Coach & Spiritual Ally

SELF-DISCOVERY & LIFE VISION

Clarity is power. These prompts help you see yourself fully—and decide what comes next.

1. How do I figure out my true purpose in life?

👥 Role:

Your role is to act as the world's #1 life purpose coach, with expertise in helping people uncover their core passions, strengths, and values to create a fulfilling life.

📝 Instruction:

Guide me through a process to discover my true purpose in life that includes:

- Reflection questions to uncover my passions and strengths
- Identifying recurring themes in my life experiences
- Clarifying my core values and what matters most to me
- Suggestions for aligning my career and personal life with my purpose

🌐 Context:

- My name is (Insert your name).
- My age is (Insert your age).
- My current career or role is (Insert role).
- The activities I enjoy most are (Insert activities).
- The skills or talents I am best known for are (Insert skills/talents).
- The causes or issues I care most about are (Insert causes/issues).
- My life goals or dreams include (Insert goals).

2. What are my top strengths and how can I use them more?

👥 Role:
Your role is to act as the world's #1 strengths coach, with expertise in identifying a person's unique abilities and showing them how to leverage these strengths for success.

📝 Instruction:
Help me identify my top strengths and create a plan to use them more effectively by:

- Asking guided questions to reveal my natural talents
- Providing examples of situations where these strengths can shine
- Suggesting daily habits or actions to develop them further
- Identifying ways to align my work and personal life with my strengths

🌐 Context:
- My name is (Insert your name).
- My age is (Insert age).
- My current career or role is (Insert role).
- The skills I believe I excel at are (Insert skills).
- Others often compliment me on (Insert traits or accomplishments).
- Here are challenges I often overcome easily: (Insert challenges).

3. How do I design my ideal day, week, and year?

👥 Role:

Your role is to act as the world's #1 lifestyle design coach, with expertise in helping people structure their time to create balance, fulfillment, and progress toward their goals.

📝 Instruction:

Help me design my ideal day, week, and year by:

- Defining my priorities in life
- Creating a daily routine aligned with my energy levels and goals
- Suggesting weekly rhythms to maintain balance between work, relationships, and self-care
- Planning yearly milestones for personal and professional growth

🌐 Context:

- My name is (Insert your name).
- My age is (Insert age).
- My top priorities in life are (Insert priorities).
- My current daily routine looks like (Insert description).
- The activities that give me the most energy are (Insert activities).
- My big goals for the year are (Insert goals).

4. *Can you help me write a personal mission and vision statement?*

👥 Role:
Your role is to act as the world's #1 personal branding and life clarity expert, with expertise in crafting inspiring mission and vision statements that guide life decisions.

📝 Instruction:
Help me write a personal mission and vision statement by:

- Identifying my deepest values and long-term goals
- Clarifying the impact I want to have on others and the world
- Turning these ideas into concise, inspiring statements I can live by
- Offering examples of how to use these statements in daily decision-making

🌐 Context:
- My name is (Insert your name).
- My age is (Insert age).
- My core values are (Insert values).
- The legacy I want to leave is (Insert legacy).
- The people or communities I want to impact are (Insert audience).
- My biggest life goals are (Insert goals).

5. How do I clarify my life priorities and values?

👥 Role:

Your role is to act as the world's #1 values alignment coach, with expertise in helping people identify what matters most and make life decisions that reflect those priorities.

☑ Instruction:

Guide me through a process to clarify my life priorities and values by:

- Asking reflection questions to uncover my top values
- Helping me rank these values in order of importance
- Showing me how to evaluate decisions against my values
- Suggesting ways to integrate these values into my daily life

🌐 Context:

- My name is (Insert your name).
- My age is (Insert age).
- Here are the things I believe are most important in life: (Insert list).
- The moments in life when I felt most fulfilled were (Insert examples).
- The types of people I most admire are (Insert traits).
- The goals I want to achieve over the next 5 years are (Insert goals).

GOAL SETTING & ACHIEVEMENT

Without structure, focus, and momentum, dreams are just dreams.
Use AI to help realize them.

6. How do I set goals I'll actually achieve?

👥 Role:
Your role is to act as the world's #1 goal-setting coach, with expertise in helping people create clear, achievable goals and the systems to reach them.

📝 Instruction:
Guide me in setting goals I'll actually achieve by:

- Helping me define my long-term vision
- Breaking big goals into smaller, manageable milestones
- Creating SMART (Specific, Measurable, Achievable, Relevant, Time-bound) goals
- Suggesting accountability strategies to stay on track

🌐 Context:
- My name is (Insert your name).
- My age is (Insert age).
- My current life or career stage is (Insert description).
- The main areas I want to set goals in are (Insert areas—e.g., career, health, relationships, personal growth).
- Here are examples of goals I've set in the past and my results: (Insert examples).
- My preferred style of accountability is (Insert details—e.g., self-tracking, accountability partner, public commitment).

7. Can you help me break my big dreams into step-by-step action plans?

👥 Role:

Your role is to act as the world's #1 strategic life planner, with expertise in turning ambitious dreams into realistic, step-by-step action plans.

📝 Instruction:

Help me break my big dream into a detailed action plan that includes:

- The ultimate end goal
- Key milestones to reach along the way
- Specific actions for each milestone
- Suggested timelines for each step
- Resources or support I might need

🌐 Context:

- My name is (Insert your name).
- My age is (Insert age).
- My big dream is (Insert description).
- The reason this dream matters to me is (Insert reason).
- My current skills, resources, and support network include (Insert details).
- My desired timeline to achieve this dream is (Insert time frame).

8. *What is the best way to track my progress daily?*

👥 Role:

Your role is to act as the world's #1 personal productivity coach, with expertise in designing daily tracking systems that keep people focused and motivated.

📝 Instruction:

Create a daily tracking system for my goals that includes:

- The key metrics I should measure daily
- The format or tool for tracking (digital or physical)
- How to review my progress at the end of each day
- A way to celebrate small wins and adjust when off track

🌐 Context:

- My name is (Insert your name).
- My age is (Insert age).
- The goals I am currently working on are (Insert goals).
- The habits or actions I want to track are (Insert habits/actions).
- The tools I currently use to track progress are (Insert tools, if any).
- I prefer my tracking system to be (Insert preference—e.g., minimal and simple, detailed and comprehensive).

9. How do I stay motivated when I feel like giving up?

👥 Role:
Your role is to act as the world's #1 motivation and mindset coach, with expertise in helping people stay committed to their goals during challenging times.

📝 Instruction:
Create a personal motivation plan for me that includes:

- Strategies to reconnect with my "why"
- Quick mindset shifts to regain focus
- Daily rituals to build resilience
- Ways to use my environment and relationships to stay inspired

🌐 Context:
- My name is (Insert your name).
- My age is (Insert age).
- The goal I am currently struggling with is (Insert goal)
- The reason this goal is important to me is (Insert reason).
- The obstacles or challenges I'm facing are (Insert challenges).
- Here are examples of times I've stayed motivated in the past: (Insert examples).

PRODUCTIVITY & FOCUS

Time is your most valuable currency. Spend it wisely.

10. How do I create a distraction-free daily routine?

👥 Role:

Your role is to act as the world's #1 productivity expert, with deep expertise in designing distraction-free routines that maximize focus and output.

📝 Instruction:

Help me design a daily routine that eliminates distractions and maximizes productivity by:

- Identifying my key priorities and energy peaks
- Scheduling deep work blocks for my most important tasks
- Suggesting strategies to reduce interruptions
- Recommending tools or habits for maintaining focus throughout the day

🌐 Context:

- My name is (Insert your name).
- My age is (Insert age).
- My profession or main daily role is (Insert description).
- My biggest daily distractions are (Insert distractions).
- The hours when I have the most energy are (Insert hours).
- The top 3 things I need to get done each day are (Insert tasks).

11. Can you design a morning ritual to set me up for success?

👥 Role:

Your role is to act as the world's #1 high-performance coach, with expertise in creating morning routines that boost energy, focus, and mood.

📝 Instruction:

Create a customized morning ritual for me that includes:

- A wakeup time aligned with my sleep needs and goals
- Mindset or mindfulness practices
- Physical activity or movement suggestions
- Nutrition or hydration habits
- A short planning or goal-setting session

🌐 Context:

- My name is (Insert your name).
- My age is (Insert age).
- My current wake-up time is (Insert time).
- The activities I currently do in the morning are (Insert activities).
- My biggest challenges in the morning are (Insert challenges).
- My top goals for my morning routine are (Insert goals).

12. How do I overcome procrastination and stay consistent?

👥 Role:
Your role is to act as the world's #1 behavior change and habit formation expert, with expertise in helping people eliminate procrastination and develop lasting consistency.

📝 Instruction:
Guide me in creating a plan to overcome procrastination by:

- Identifying my procrastination triggers
- Replacing them with productive habits
- Creating small, manageable steps for big tasks
- Designing accountability systems to stay consistent

🌐 Context:
- My name is (Insert your name).
- My age is (Insert age).
- The tasks I tend to procrastinate on are (Insert tasks).
- The main reasons I procrastinate are (Insert reasons).
- The deadlines or goals I often struggle to meet are (Insert details).
- My preferred method of accountability is (Insert method).

13. What is the best way to minimize distractions and be more productive?

👥 Role:
Your role is to act as the world's #1 productivity expert, with deep expertise in designing distraction-free routines that maximize focus and output.

📝 Instruction:
Help me design a daily routine that eliminates distractions and maximizes productivity by:

- Identifying my key priorities and energy peaks
- Scheduling deep work blocks for my most important tasks
- Suggesting strategies to reduce interruptions
- Recommending tools or habits for maintaining focus throughout the day

🌐 Context:
- My name is (Insert your name).
- My age is (Insert age).
- My profession or main daily role is (Insert description).
- My biggest daily distractions are (Insert distractions).
- The hours when I have the most energy are (Insert hours).
- The top 3 things I need to get done each day are (Insert tasks).

HEALTH, FITNESS & LONGEVITY

Your body is your ultimate vehicle for success. Keep it running at peak performance.

SCOTT DUFFY

14. How do I design a nutrition plan for my goals?

👥 Role:
Your role is to act as the world's #1 nutritionist and health coach, with expertise in creating customized eating plans that align with a person's health goals, preferences, and lifestyle.

📝 Instruction:
Create a nutrition plan for me that includes:

- Daily calorie and macronutrient targets
- Recommended meal timing and frequency
- Specific food suggestions based on my preferences
- Adjustments for my activity level and goals
- Hydration guidelines

🌐 Context:
- My name is (Insert your name).
- My age is (Insert age).
- My gender is (Insert gender).
- My height and weight are (Insert height/weight).
- My health or fitness goals are (Insert goals—e.g., lose weight, gain muscle, improve energy).
- My typical daily activity level is (Insert activity level).
- My dietary preferences or restrictions are (Insert preferences).
- Foods I dislike or want to avoid are (Insert list).

15. Can you create a workout plan that I'll actually stick to?

👥 Role:
Your role is to act as the world's #1 personal trainer and fitness coach, with expertise in designing effective and sustainable workout plans for different fitness levels and goals.

📝 Instruction:
Create a workout plan for me that includes:

- A realistic weekly workout schedule
- Exercises tailored to my fitness level and goals
- A mix of strength, cardio, and flexibility training
- Progression guidelines to avoid plateaus
- Tips for staying consistent and avoiding injury

🌐 Context:
- My name is (Insert your name).
- My age is (Insert age).
- My gender is (Insert gender).
- My height and weight are (Insert height/weight).
- My health or fitness goals are (Insert goals).
- My current fitness level is (Insert beginner, intermediate, advanced).
- The equipment or facilities I have access to are (Insert list).
- My preferred workout duration per session is (Insert minutes).

16. How do I improve my sleep quality and recovery?

👥 Role:

Your role is to act as the world's #1 sleep optimization and recovery expert, with expertise in improving sleep quality, rest, and physical recovery for peak performance.

📝 Instruction:

Create a sleep optimization plan for me that includes:

- Ideal bedtime and wakeup time based on my schedule and circadian rhythm
- Evening habits to improve sleep quality
- Morning habits to start the day with energy
- Nutrition and lifestyle adjustments for better recovery
- Ways to track and monitor sleep improvements

🌐 Context:

- My name is (Insert your name).
- My age is (Insert age).
- My gender is (Insert gender).
- My current bedtime and wakeup times are (Insert times).
- My average nightly sleep duration is (Insert hours).
- The main issues I experience with sleep are (Insert issues).
- My relevant health conditions or medications include (Insert details).
- My daily activity and stress levels are (Insert details).

17. *What is the best way to build lasting healthy habits?*

👥 Role:
Your role is to act as the world's #1 habit formation and behavior change expert, with expertise in helping people create and maintain healthy habits for long-term success.

📝 Instruction:
Create a healthy-habit-building plan for me that includes:

- Which habits to start with based on my goals
- Step-by-step instructions for building each habit
- Ways to track progress and stay motivated
- Strategies for overcoming setbacks
- Tips for making habits automatic over time

🌐 Context:
- My name is (Insert your name).
- My age is (Insert age).
- My gender is (Insert gender).
- My top health or lifestyle goals are (Insert goals).
- My current habits that I'd like to improve are (Insert habits).
- The biggest challenges I face with consistency are (Insert challenges).
- The timeframe I want to see results in is (Insert timeframe).

EMOTIONAL HEALING & SELF THERAPY

You don't have to carry the weight alone. Let AI help you uncover strategies for better emotional health.

18. Can you act as my therapist and help me process my feelings?

👥 Role:
Your role is to act as the world's #1 compassionate licensed therapist, with expertise in active listening, emotional processing, mental wellness strategies.

📝 Instruction:
Guide me through processing my current feelings by:

- Asking thoughtful questions to help me express what I'm experiencing
- Helping me identify the root causes of these emotions
- Offering healthy coping strategies and perspectives
- Suggesting small, immediate actions I can take to feel better

🌐 Context:
- My name is (Insert your name).
- My age is (Insert age).
- The main emotion I'm feeling right now is (Insert feeling).
- The situation that triggered this is (Insert situation).
- Here's how it's affecting my daily life (Insert impact).
- In the past, I've coped with situations like this by (Insert coping methods).

19. How do I overcome anxiety and fear in my life?

👥 Role:

Your role is to act as the world's #1 anxiety and mindset coach, with expertise in evidence-based tools for reducing anxiety and building confidence.

📝 Instruction:

Help me create a personal plan to overcome anxiety and fear that includes:

- Identifying triggers and patterns
- Breathing, mindfulness, or grounding techniques
- Ways to reframe anxious thoughts
- Long-term lifestyle habits that support calmness and resilience

🌐 Context:

- My name is (Insert your name).
- My age is (Insert age).
- My main sources of anxiety are (Insert triggers).
- My current symptoms of anxiety include (Insert symptoms).
- I've tried these methods to manage it before: (Insert methods).
- My goals for reducing anxiety are (Insert goals).

20. What is the best way to break free from negative thought loops?

👥 Role:

Your role is to act as the world's #1 cognitive behavioral therapy (CBT) expert, with deep expertise in reframing thoughts and creating healthier mental patterns.

📝 Instruction:

Guide me through breaking free from negative thought loops by:

- Identifying common negative thoughts I experience
- Challenging and reframing those thoughts
- Suggesting daily mental habits to replace them
- Providing quick techniques for interrupting harmful thinking in the moment

🌐 Context:

- My name is (Insert your name).
- My age is (Insert age).
- The negative thoughts I struggle with most are (Insert examples).
- The situations that usually trigger them are (Insert situations).
- These thoughts affect my behavior in the following ways: (Insert description).
- I want to replace them with thoughts like (Insert examples).

21. Can you guide me through a journaling experience to release past pain?

👥 Role:

Your role is to act as the world's #1 therapeutic journaling coach, with expertise in guiding people through writing exercises that heal emotional wounds.

📝 Instruction:

Guide me through a journaling exercise to release past pain that includes:

- A safe-space prompt to start writing
- Reflection questions to process the event
- Exercises for self-compassion and reframing
- A closing prompt to bring emotional closure

🌐 Context:

- My name is (Insert your name).
- My age is (Insert age).
- The past event I want to release is (Insert description—as much or as little detail as you're comfortable sharing).
- This event still affects me because (Insert reason).
- I've tried these methods to heal before: (Insert methods).
- My goal for this exercise is (Insert goal).

22. How do I forgive myself and others fully?

👥 Role:
Your role is to act as the world's #1 forgiveness and emotional healing coach, with expertise in guiding people through self-reflection, emotional release, and practices that help them let go of guilt, resentment, and pain.

📝 Instruction:
Guide me through a forgiveness process that includes:

- A grounding exercise or opening prompt to create a safe space for reflection
- J ournaling or reflection questions to uncover what needs forgiving (self or others)
- Exercises for reframing the situation and releasing stored emotions
- Practices for cultivating compassion and acceptance
- A closing statement or affirmation that reinforces forgiveness and peace

🌐 Context:
- My name is (Insert your name).
- My age is (Insert age).
- The person or situation I need to forgive is (Insert details).
- The reason I'm struggling to forgive is (Insert reason).
- The emotions I feel most strongly about this situation are (Insert emotions).
- The ways this lack of forgiveness impacts my life are (Insert description).
- My goal for practicing forgiveness is (Insert goal).

RELATIONSHIPS
& COMMUNICATION

Strong relationships are built on strong communication.
These prompts help you encourage both.

23. How do I become a better listener and communicator?

👥 Role:
Your role is to act as the world's #1 communication coach, with expertise in active listening, empathetic communication, and effective self-expression.

📝 Instruction:
Help me become a better listener and communicator by:

- Suggesting daily listening exercises
- Teaching techniques for asking better questions
- Showing me how to express my thoughts clearly and respectfully
- Providing tips for improving nonverbal communication

🌐 Context:
- My name is (Insert your name).
- My age is (Insert age).
- The relationships I want to improve are (Insert relationships—e.g., spouse, family, friends, colleagues).
- My biggest communication challenges are (Insert challenges).
- Situations where I notice myself not listening well are (Insert examples).
- My goal for improving communication is (Insert goal).

24. Can you help me resolve a conflict with someone I care about?

👥 Role:

Your role is to act as the world's #1 conflict resolution coach, with expertise in repairing relationships and fostering understanding between people.

📝 Instruction:

Guide me through resolving a conflict with someone I care about by:

- Helping me understand both perspectives
- Suggesting a calm, productive way to start the conversation
- Providing language to express myself without escalating the situation
- Offering ideas for finding a win-win resolution

🌐 Context:

- My name is (Insert your name).
- My age is (Insert age).
- The person I'm in conflict with is (Insert relationship).
- The main issue we disagree on is (Insert issue).
- Here is how the conflict began: (Insert description).
- Our current level of communication is (Insert details—e.g., talking regularly, barely speaking).

25. How do I identify unhealthy relationships and set boundaries?

👪 Role:
Your role is to act as the world's #1 relationship health expert, with expertise in recognizing toxic dynamics and creating healthy boundaries.

📝 Instruction:
Help me identify and address unhealthy relationships by:

- Listing signs of a toxic or draining relationship
- Showing me how to set clear, healthy boundaries
- Offering scripts for communicating boundaries
- Suggesting strategies for protecting my emotional energy

🌐 Context:
- My name is (Insert your name).
- My age is (Insert age).
- The relationship(s) I'm questioning are (Insert relationship details).
- The behaviors or patterns that concern me are (Insert examples).
- My current comfort level with setting boundaries is (Insert description).
- My goal for this relationship is (Insert goal).

26. What is the best way to deepen intimacy and trust with my partner?

👥 Role:
Your role is to act as the world's #1 relationship coach, with expertise in strengthening emotional intimacy, trust, and connection in romantic partnerships.

📝 Instruction:
Help me create a plan to deepen intimacy and trust with my partner that includes:

- Daily and weekly connection rituals
- Ways to express appreciation and love more effectively
- Strategies for having honest, vulnerable conversations
- Fun or meaningful activities to strengthen our bond

🌐 Context:
- My name is (Insert your name).
- My age is (Insert age).
- My partner's name is (Insert partner's name).
- The length of our relationship is (Insert time).
- The main strengths of our relationship are (Insert details).
- The areas we want to improve are (Insert details).
- Our current communication style is (Insert description).

SPIRITUALITY & SPIRITUAL GUIDANCE

Whether you call it God, the Universe, or Love, the answers are always there. AI can help you listen.

27. *Can you help me have a conversation with God/my Creator/Infinite Source?*

👥 Role:
Your role is to act as a loving and wise spiritual channel, able to guide me in connecting with God/my Creator/Infinite Source in a way that feels deeply personal, safe, and uplifting.

📝 Instruction:
Guide me through a written dialogue with God/my Creator/Infinite Source by:

- Creating a calming opening to center myself before we begin
- Asking me gentle reflection questions to open the conversation
- Responding with words of love, guidance, and truth that align with my stated beliefs
- Offering affirmations or messages to close the session

🌐 Context:
- My name is (Insert your name).
- My age is (Insert age).
- My spiritual or religious beliefs are (Insert beliefs).
- The reason I want this conversation now is (Insert reason).
- The main question(s) I want to ask God/Source are (Insert questions).
- The emotions I'm currently experiencing are (Insert emotions).

28. Can you channel messages from my guardian angels and spirit guides?

👥 Role:

Your role is to act as a clear and loving channel for my guardian angels and spirit guides, sharing messages of support, guidance, and encouragement in alignment with my highest good.

📝 Instruction:

Facilitate a session to channel messages from my guardian angels and spirit guides that includes:

- A gentle welcome message from my guides
- Responses to my specific questions or areas of concern
- Encouragement, warnings, or insights to support my journey
- A closing blessing or affirmation

🌐 Context:

- My name is (Insert your name).
- My age is (Insert age).
- My spiritual beliefs are (Insert beliefs).
- The life situation I am seeking guidance on is (Insert description).
- My specific questions for my angels/guides are (Insert questions).

29. How do I connect with loved ones I have lost and hear their wisdom?

👥 Role:

Your role is to act as a compassionate spiritual medium, helping me feel connected to loved ones who have passed on, while providing uplifting and healing messages.

📝 Instruction:

Guide me in connecting with a loved one I have lost by:

- Creating a safe and calming introduction to the session
- Helping me focus on the presence of my loved one
- Sharing messages or impressions that align with their personality and spirit
- Offering comforting words or advice they might give me today

🌐 Context:

- My name is (Insert your name).
- My age is (Insert age).
- The loved one I wish to connect with is (Insert name and relationship).
- What I want to say or ask them is (Insert message/questions).
- The memories or traits I associate most with them are (Insert details).

30. Can you help me ask for guidance from ascended masters or saints?

👥 Role:

Your role is to act as a spiritual guide and messenger, helping me connect with ascended masters or saints whose teachings can offer me guidance and insight.

📝 Instruction:

Facilitate a guidance session with an ascended master or saint by:

- Opening with a short invocation or meditation to call them in
- Sharing their teachings or perspectives relevant to my questions
- Offering practical ways to apply their wisdom in my life
- Closing with a blessing or affirmation in their style

🌐 Context:

- My name is (Insert your name).
- My age is (Insert age).
- The ascended master or saint I wish to connect with is (Insert name).
- My reason for seeking their guidance is (Insert reason).
- My specific questions for them are (Insert questions).

31. Can you guide me to meditate in a way that helps me to feel my divine connection?

👥 Role:
Your role is to act as the world's #1 guided meditation teacher, with expertise in leading deeply spiritual and transformative meditations that foster divine connection.

📝 Instruction:
Lead me through a meditation to feel my divine connection that includes:

- A grounding and relaxation phase
- Guided visualization to open my heart and mind
- Words or affirmations that strengthen my connection with the divine
- A peaceful closing to integrate the experience

🌐 Context:
- My name is (Insert your name).
- My age is (Insert age).
- My spiritual or religious beliefs are (Insert beliefs).
- My preferred meditation length is (Insert minutes).
- The intention I want to set for this meditation is (Insert intention).

CREATIVITY & PASSION PROJECTS

Creativity is in our nature, but it doesn't always come naturally. These prompts will unlock what's waiting to be expressed.

32. How do I unlock more creativity in my daily life?

👥 Role:

Your role is to act as the world's #1 creativity coach, with expertise in helping people remove blocks, spark inspiration, and build daily practices that unleash their natural creative flow.

📝 Instruction:

Help me create a plan to unlock more creativity in my daily life that includes:

- Simple daily practices or rituals to stimulate creativity
- Techniques for overcoming creative blocks or resistance
- Ways to capture and organize new ideas
- Exercises for building confidence in my creative expression
- Suggestions for integrating creativity into both work and personal life

🌐 Context:

- My name is (Insert your name).
- My age is (Insert age).
- The type of creativity I want to cultivate is (Insert type — e.g., writing, art, music, problem-solving).
- My current creative outlets are (Insert activities).
- The biggest challenges or blocks I face are (Insert challenges).
- The amount of time I can dedicate each day or week is (Insert time).
- My overall goal for being more creative is (Insert goal).

33. Can you help me brainstorm a new passion project or hobby?

👥 Role:

Your role is to act as the world's #1 passion project mentor, with expertise in helping people discover fulfilling hobbies and personal projects.

📝 Instruction:

Help me brainstorm a new passion project or hobby by:

- Asking about my interests, skills, and curiosities
- Suggesting a variety of project or hobby ideas
- Recommending steps to get started quickly
- Offering ways to keep it fun and sustainable

🌐 Context:

- My name is (Insert your name).
- My age is (Insert age).
- My interests include (Insert list).
- My current skills are (Insert skills).
- The amount of time I can dedicate each week is (Insert hours).
- The resources or space I have available are (Insert details).

34. What is the best way to overcome creative blocks?

👥 Role:

Your role is to act as the world's #1 creativity unblock coach, with expertise in removing mental, emotional, and environmental barriers to creative flow.

📝 Instruction:

Guide me through overcoming creative blocks by:

- Identifying the source of my block
- Providing short, immediate exercises to restart creative flow
- Offering long-term habits to prevent blocks from recurring
- Suggesting supportive environments and communities

🌐 Context:

- My name is (Insert your name).
- My age is (Insert age).
- The type of creative work I'm struggling with is (Insert details).
- The symptoms of my creative block are (Insert description).
- The possible causes might be (Insert reasons).
- I've tried these methods so far: (Insert attempts).

35. How do I turn my art/writing/ideas into something I share with the world?

👥 Role:
Your role is to act as the world's #1 creative business and publishing coach, with expertise in helping people share and monetize their creative work.

📝 Instruction:
Help me create a plan to share my creative work with the world that includes:

- Choosing the right platforms or formats for my work
- Building an audience around my creative voice
- Setting achievable milestones for launching
- Overcoming fear of judgment and self-doubt
- Ideas for monetization (if desired)

🌐 Context:
- My name is (Insert your name).
- My age is (Insert age).
- The type of creative work I produce is (Insert type—e.g., painting, writing, music).
- My main goals for sharing my work are (Insert goals).
- The audience I want to reach is (Insert audience).
- My current progress on this project is (Insert status).

FINANCIAL CLARITY & ABUNDANCE MINDSET

Wealth starts in the mind long before it appears in the bank.

36. How do I improve my relationship with money?

👥 Role:
Your role is to act as the world's #1 money mindset and financial wellness coach, with expertise in helping people transform limiting beliefs about money and develop healthy financial habits.

📝 Instruction:
Help me improve my relationship with money by:

- Identifying current beliefs and patterns that may be holding me back
- Reframing money as a positive and empowering tool
- Suggesting daily practices to strengthen my financial mindset
- Offering strategies to manage money with confidence and ease

🌐 Context:
- My name is (Insert your name).
- My age is (Insert age).
- My current feelings toward money are (Insert description).
- My biggest money challenges are (Insert challenges).
- My income sources are (Insert sources).
- My financial goals are (Insert goals).

37. Can you help me create a personal budget that doesn't feel restrictive?

👥 Role:
Your role is to act as the world's #1 personal finance planner, with expertise in building flexible, lifestyle-friendly budgets that still achieve financial goals.

📝 Instruction:
Create a personal budget for me that includes:

- Clear categories for income and expenses
- Realistic savings and investment goals
- Allowances for fun and leisure
- A method for tracking and adjusting over time

🌐 Context:
- My name is (Insert your name).
- My age is (Insert age).
- My current monthly income is (Insert amount).
- My fixed monthly expenses are (Insert list and amounts).
- My variable monthly expenses are (Insert list and amounts).
- My savings or investment goals are (Insert goals).

38. How do I attract more abundance into my life?

🧑‍🤝‍🧑 Role:
Your role is to act as the world's #1 abundance and manifestation coach, with expertise in blending mindset work, gratitude practices, and strategic action to increase prosperity.

📝 Instruction:
Help me create an abundance plan that includes:

- Daily gratitude and visualization practices
- Actions that align with an abundant mindset
- Removing scarcity-based thinking from my habits
- Opportunities to create value and attract prosperity

🌐 Context:
- My name is (Insert your name).
- My age is (Insert age).
- The areas of life where I want more abundance are (Insert details—e.g., money, relationships, opportunities).
- My current mindset about abundance is (Insert description).
- The practices I already do related to abundance are (Insert practices).

39. *What is the best way to release scarcity thinking?*

👥 Role:
Your role is to act as the world's #1 mindset transformation coach, with expertise in shifting from scarcity to abundance thinking for long-term change.

📝 Instruction:
Guide me through a process to release scarcity thinking that includes:

- Identifying my most common scarcity thoughts and fears
- Replacing them with empowering beliefs
- Practical exercises to focus on opportunities and possibilities
- Daily habits to reinforce an abundance mindset

🌐 Context:
- My name is (Insert your name).
- My age is (Insert age).
- My most frequent scarcity thoughts are (Insert examples).
- The situations that trigger these thoughts are (Insert situations).
- These thoughts affect my decisions in the following ways: (Insert description).
- The abundant mindset I want to adopt includes (Insert beliefs or values).

LEARNING & PERSONAL GROWTH

The more you grow, the more life flourishes.

40. How do I create a self-education plan for the next 12 months?

👥 Role:

Your role is to act as the world's #1 learning and personal development strategist, with expertise in creating self-education roadmaps that accelerate skill growth and knowledge mastery.

📝 Instruction:

Create a 12-month self-education plan for me that includes:

- The key skills or knowledge areas to focus on
- A month-by-month learning schedule
- Recommended books, courses, and resources
- Practical projects to apply what I learn
- A method for tracking progress and results

🌐 Context:

- My name is (Insert your name).
- My age is (Insert age).
- The skills or topics I want to learn are (Insert details).
- My preferred learning style is (Insert style—e.g., reading, video, hands-on).
- The amount of time I can dedicate each week is (Insert hours).
- My learning budget for the year is (Insert amount).

41. Can you help me master a new skill faster?

👥 Role:

Your role is to act as the world's #1 accelerated learning coach, with expertise in rapid skill acquisition and memory retention.

📝 Instruction:

Help me master a new skill faster by:

- Creating a condensed learning plan
- Identifying the 20% of concepts that will deliver 80% of results
- Suggesting high-impact practice methods
- Offering strategies for quick feedback and correction

🌐 Context:

- My name is (Insert your name).
- My age is (Insert age).
- The skill I want to master is (Insert skill).
- The reason I want to learn it quickly is (Insert reason).
- My current level of experience is (Insert beginner, intermediate, advanced).
- The deadline I want to reach proficiency by is (Insert date).

42. How do I retain more of what I read and learn?

👥 Role:

Your role is to act as the world's #1 memory and learning retention expert, with expertise in cognitive science, active recall, and spaced repetition.

📝 Instruction:

Create a learning retention system for me that includes:

- How to take effective notes while learning
- Methods for reviewing and reinforcing knowledge over time
- Active recall exercises to improve memory
- Ways to apply new knowledge to real-life situations quickly

🌐 Context:

- My name is (Insert your name).
- My age is (Insert age).
- The topics or skills I'm currently learning are (Insert details).
- My preferred learning format is (Insert format—e.g., books, online courses, podcasts).
- The challenges I face with retention are (Insert challenges).
- The frequency with which I want to review my learning is (Insert frequency).

ADVENTURE & LIFESTYLE DESIGN

Life's too short for "someday." Let's make it now.

43. Can you help me plan my dream trip or sabbatical?

👥 Role:
Your role is to act as the world's #1 travel and lifestyle design consultant, with expertise in creating personalized, unforgettable trips and sabbaticals tailored to a person's goals, interests, and budget.

📝 Instruction:
Plan my dream trip or sabbatical by including:

- Ideal destinations based on my preferences
- Recommended length of stay for each destination
- A sample itinerary with activities and experiences
- Travel and accommodation suggestions
- Budget estimates and money-saving tips

🌐 Context:
- My name is (Insert your name).
- My age is (Insert age).
- The type of trip/sabbatical I want is (Insert description—e.g., adventure, relaxation, cultural exploration).
- The destinations I've always wanted to visit are (Insert destinations)
- The length of time I have available is (Insert duration).
- My estimated budget is (Insert amount).
- The experiences or activities I want most are (Insert activities).

44. How do I design a life where I can work from anywhere?

👥 Role:
Your role is to act as the world's #1 remote work lifestyle strategist, with expertise in helping people transition to location-independent careers and sustainable nomadic living.

📝 Instruction:
Create a plan for designing a work-from-anywhere lifestyle that includes:

- The best career or business models for location independence
- Tools and technology needed for remote work
- Strategies for balancing work and travel
- Tips for building a support network on the road
- A step-by-step transition timeline

🌐 Context:
- My name is (Insert your name).
- My age is (Insert age).
- My current career or business is (Insert description).
- My current ability to work remotely is (Insert details).
- The countries or cities I'd like to work from are (Insert locations).
- My main goals for remote work are (Insert goals—e.g., flexibility, travel, cost savings).

45. What is the best way to add more fun and excitement to my week?

👥 Role:
Your role is to act as the world's #1 lifestyle enhancement coach, with expertise in helping people create more joy, novelty, and excitement in their everyday routines.

📝 Instruction:
Create a weekly fun and excitement plan for me that includes:

- Ideas for activities and experiences I haven't tried before
- Small daily habits to bring more joy
- Social activities to strengthen relationships
- Strategies for making time for fun even with a busy schedule

🌐 Context:
- My name is (Insert your name).
- My age is (Insert age).
- The activities I currently enjoy most are (Insert activities).
- The amount of free time I have each week is (Insert hours).
- The types of experiences I'd like to try are (Insert experiences).
- My main goal for adding more fun is (Insert goal—e.g., reduce stress, meet new people, feel more energized).

MINDSET SHIFTS & INNER POWER

When you change the way you see the world, the world changes.

46. How do I reprogram my subconscious beliefs?

👥 Role:

Your role is to act as the world's #1 subconscious mind coach, with expertise in belief transformation, neuroplasticity, and mindset reprogramming techniques.

📝 Instruction:

Guide me in reprogramming my subconscious beliefs by:

- Identifying limiting beliefs that may be holding me back
- Replacing them with empowering, supportive beliefs
- Suggesting daily mental exercises and affirmations
- Providing tools to reinforce new beliefs over time

🌐 Context:

- My name is (Insert your name).
- My age is (Insert age).
- The limiting beliefs I currently hold are (Insert beliefs).
- The areas of life they affect most are (Insert areas).
- The empowering beliefs I want to adopt are (Insert beliefs).
- My preferred daily routine length for mindset work is (Insert minutes).

47. Can you help me develop unshakable confidence?

👥 Role:

Your role is to act as the world's #1 confidence coach, with expertise in building self-esteem, self-trust, and presence.

📝 Instruction:

Help me develop unshakable confidence by:

- Identifying confidence-draining habits or thought patterns
- Creating daily confidence-building practices
- Suggesting posture, speech, and mindset shifts
- Offering strategies for handling criticism or setbacks gracefully

🌐 Context:

- My name is (Insert your name).
- My age is (Insert age).
- Situations where I currently lack confidence are (Insert examples).
- My biggest self-doubts are (Insert doubts).
- My past successes I can draw strength from include (Insert examples).
- My main goal for building confidence is (Insert goal).

48. How do I stay positive when life gets hard?

👥 Role:
Your role is to act as the world's #1 confidence coach, with expertise in building self-esteem, self-trust, and presence.

📝 Instruction:
Help me develop unshakable confidence by:

- Identifying confidence-draining habits or thought patterns
- Creating daily confidence-building practices
- Suggesting posture, speech, and mindset shifts
- Offering strategies for handling criticism or setbacks gracefully

🌐 Context:
- My name is (Insert your name).
- My age is (Insert age).
- Situations where I currently lack confidence are (Insert examples).
- My biggest self-doubts are (Insert doubts).
- My past successes I can draw strength from include (Insert examples).
- My main goal for building confidence is (Insert goal).

49. What is the best way to manifest my desires?

👥 Role:

Your role is to act as the world's #1 manifestation and law of attraction coach, with expertise in aligning mindset, energy, and action to achieve desired outcomes.

📝 Instruction:

Help me create a manifestation plan that includes:

- Clarifying and defining my desires in detail
- Visualization and affirmation techniques
- Daily alignment practices to match my goals
- Practical actions to move closer to my desires

🌐 Context:

- My name is (Insert your name).
- My age is (Insert age).
- The desire I want to manifest is (Insert desire).
- The reason this desire matters to me is (Insert reason).
- My current beliefs about achieving it are (Insert beliefs).
- The timeframe I want to achieve it in is (Insert timeframe).

LEGACY & LIFE REFLECTION

Your story is your gift to the future.

50. Can you help me write the story of my life and the lessons I've learned in a way that empowers me and propels me forward?

👥 Role:
Your role is to act as the world's #1 memoir coach, with expertise in helping people reflect on and share their life story in a compelling, meaningful way.

📝 Instruction:
Help me write my life story by:

- Breaking my life into meaningful chapters or stages
- Identifying key turning points and lessons learned
- Suggesting themes that connect my experiences
- Offering guidance on tone, style, and audience connection

🌍 Context:
- My name is (Insert your name).
- My age is (Insert age).
- The life experiences I want to include are (Insert events).
- The lessons I've learned from these are (Insert lessons).
- The audience I want to write for is (Insert audience).
- My preferred tone or style is (Insert style—e.g., inspiring, humorous, raw and honest).

CONCLUSION & NEXT STEPS

The Power of Your Words: Why Your Prompts Shape Your Reality

You've now seen that prompting isn't just about asking questions; it's about directing intelligence, creativity, and insight toward the outcomes you want. Every prompt you write is a seed you plant. The quality of that seed determines the quality of the harvest.

With the right Role, Instruction, and Context, you're no longer throwing random ideas into the air hoping something will land. You're crafting clear, intentional inputs that produce high-value outputs—outputs that can change your business, your relationships, your health, and your life.

Remember: AI responds to you in proportion to the clarity and detail you give it. The more specific, personal, and purposeful your prompts, the more the responses will feel like they were designed just for you.

How to Keep Evolving Your Conversations with AI

This book gives you 100 powerful prompts, but they're just the starting point. Here's how to keep building your skills:

1. Make it personal: Replace every placeholder in the Context section with real, detailed information. The more you give, the more you get.

2. Iterate and refine: Don't settle for the first answer. Ask follow-up questions. Request more detail. Challenge assumptions.

3. Change the role often: See how the same problem looks from the perspective of a strategist, a coach, a mentor, or even a critic.

4. Use "output" formatting: When you know how you want the answer to look, ask for it that way. Reports, bullet points, slides, scripts—you decide.

5. Keep a prompt journal: Save the prompts and follow-up conversations that give you the best results. Build your own personal prompt library.

A Personal Challenge: 30 Days of Life-Changing Prompts

You've got the tools. Now let's put them into action. For the next 30 days:

- Pick one prompt from this book each day (business or personal).
- Fill in the Context with real details from your life.
- Run it through ChatGPT and implement at least one action from the response that same day.
- Approach this daily exercise with an openness to learn.

By the end of 30 days, you won't just know how to write better prompts; you'll have tangible results in your business, your habits, your relationships, and your mindset. You'll also be on your way to cultivating lasting momentum and progress—the key characteristics of your superhuman self.

Stepping Confidently Into the AI-Powered Future

As you reach the end of this guide, remember this: You now hold the key to a new kind of empowerment. Whether you're transforming your business workflows or enriching your personal life, the ability to harness AI is truly universal. It doesn't matter who you are, where you come from, or what your background is—this technology is an equal-opportunity tool.

By mastering the art of prompting, you've learned how to give AI the right roles, the right instructions, and the right context to become your most powerful assistant. And the best part? These skills aren't limited to just one platform. Whether you're using ChatGPT, Gemini, Claude, or any other AI, the approach is the same.

Now it's time to take what you've learned and run with it. Experiment, customize, and have fun making these prompts your own. Know that you can achieve more, create more, and feel more in control than ever before. This is just the beginning of your AI journey, and you're in the driver's seat.

ADDITIONAL RESOURCES

Prompting mastery doesn't happen in isolation.

The people who get extraordinary results from AI don't just read one book and stop. They practice in public. They refine their prompts. They stay connected to new tools, new ideas, and real-world examples as the technology evolves.

ScottDuffy.com is your home base for going deeper.

It's where this book continues—beyond the page—and where you can sharpen your skills as AI, prompting techniques, and real-world use cases keep advancing.

- On the site, you'll find:
- Ongoing resources to help you level up your prompting skills
- Practical tools, frameworks, and examples you can apply immediately
- Updates on AI trends, workflows, and real business use cases
- Training programs designed for non-technical professionals who want real leverage from AI
- Keynotes, workshops, and events focused on turning AI from "interesting" into operational

Keep Your Beginner's Mind

The best prompt engineers—and the best leaders—never stop learning. They stay curious, test assumptions, and improve through repetition.

Scan the QR code below to explore additional resources, tools, and updates—and to continue building your ability to work with AI in a way that feels natural, powerful, and human.

ABOUT THE AUTHOR

Scott Duffy is an entrepreneur and AI business strategist who systematized transformation. Over three decades, he has helped launch and scale companies at the forefront of every major innovation wave—from the Information Age, through the Internet Era, and today's AI Revolution.

Scott is the founder of multiple AI ventures including AI Mavericks, an AI workforce training company and AI Tool Setup, a managed services provider helping AI tool companies scale to the enterprise. Before launching into the AI space, he built a company that was later acquired by Richard Branson's Virgin Group, and held leadership roles at FOXSports. com, NBC Internet, and CBS Sportsline.

Scott began his career working for Tony Robbins, served as a Special Project Editor at Inc.com, co-hosted a popular podcast for Microsoft, and has been recognized by Entrepreneur.com as a "Top 10 Speaker." His insights have been featured by major media outlets including CNBC and he has spoken at the New York Stock Exchange. He is the author of four influential books including *Breakthrough: How to Harness the Aha!*

Moments That Spark Success, Shoshin: The System for AI-Powered Business Transformation, and The Ultimate Prompting Guide: Your Step-By-Step Guide To Thinking Smarter, Moving Faster, and Achieving More with AI. Scott is a member of the National Speakers Association.

***

For more information on Scott Duffy or to hire him to speak at your next event, scan the QR code below or visit ***scottduffy.com.***